The Triangle Triplets

by Mary E. Gentry
Illustrated by Elise Whittier Church

ISBN: 978-1-944613-36-5

Once upon a time,
Mr. and Mrs. Triangle had triplets.
As you can imagine, they were very
excited about their babies.

Soon it was time
to think of names for their
three little bundles of joy!

But deciding what to call them was harder than they imagined.

So they waited and watched for ideas of what they could name their newborns.

The perfect name for the
first little triangle soon became apparent,
as she loved to
share her snacks equally
between her two brothers and herself.

She could often be heard settling arguments by saying, "One for you, one for you, and one for me. Now we're equal!"

So Mr. and Mrs. Triangle named her Equilateral Triangle.

As she became older, Equilateral's personality and shape took on her name.

No matter how much she grew, each of her three sides was always equal.

In other words, each side was the same length.

Well, the name for the second of the three triangle triplets was just as hard to decide upon. You see, triplet two was totally different than Equilateral, as none of his sides were the same length.

In fact, each of his sides was a different length. How he got his name is also a unique story.

When triplet two was just a few weeks old, it was time for Mr. and Mrs. Triangle to take the babies to the doctor for a check-up.

On that particular day, triplet two would not leave any of his favorite playthings at home.

So when it was his turn to be weighed on the scale, the weight of all his toys made him lean to the side. This caused his shape to stretch and all his sides to become different lengths.

One side was a short length, one side was a medium length, and the third side was a very long length. Right then and there, Mr. and Mrs. Triangle decided to name their second baby Scalene Triangle, as it was while he was leaning on the scale they determined his name.

Now the last triplet in the Triangle family still needed his own name. After all, a triangle can't go around his or her entire life without a name!

For Mr. and Mrs. Triangle, selecting a name for the third triangle triplet was exceptionally difficult. But eventually an idea came. You guessed it! Another interesting story!

You see, every day triangle three would run inside the house and yell, "Guess what I saw! Guess what I saw!"

He had a habit of peeking over the neighbor's fence to see into the neighbor's garden.

Some days he saw fluffy white rabbits eating the carrots.

Other days he saw beautiful yellow butterflies hovering above the pink daisy bushes.

No matter what he spotted, he always wanted to share the news by shouting, "Guess what I saw! Guess what I saw!"

With all that stretching and peeking and peeking and stretching, triangle three started growing taller and taller and taller.

Taking a long look at their triplet, who now had two extra long sides of the same length, Mr. and Mrs. Triangle decided to call their third triplet Isosceles Triangle. It's pronounced "I-SAW-sih-leez". Say hello to baby Isosceles.

And now you know how the triangle triplets got their names!

But wait. . . I forgot to tell you about the triangle family who lived next door. . .

Mr. and Mrs. Right!

And that my friends, is another story.

Create your own triangles, kids!

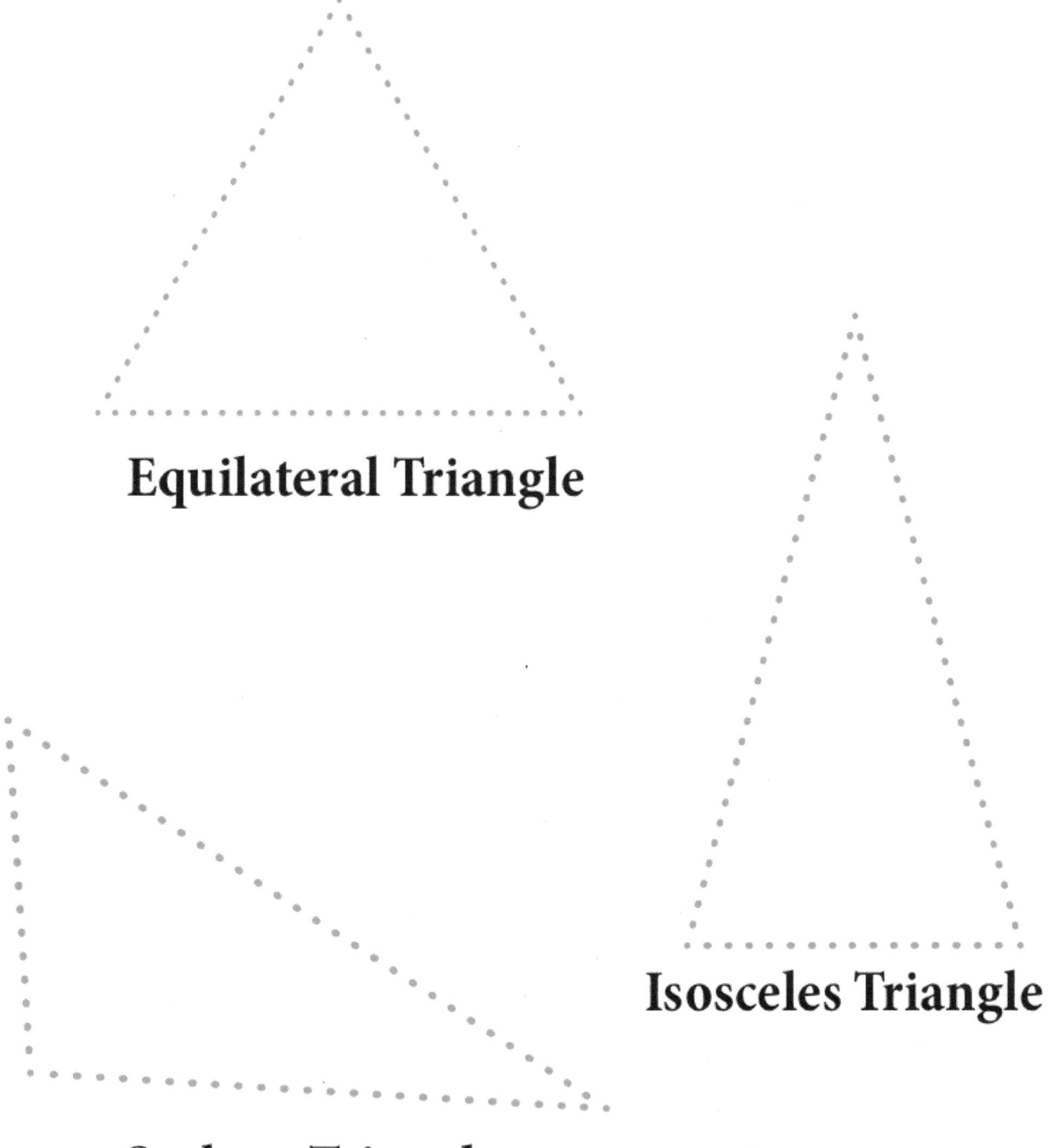

Scalene Triangle

MARY E. GENTRY has taken an ordinary math standard and transformed it into a "Word Picture Story" that will forever cement the triangle triplets skill into the minds and memories of young readers. While teaching over twenty-six years in California and Oklahoma, Mary once saved a baby duckling's life by scrubbing it from its stuck position with a toothbrush. She has also jumped out of a perfectly good airplane to experience the thrill of flying! When not grading papers, she can be found riding bikes with her husband or enjoying popcorn and M & Ms at the movies.

CPSIA information can be obtained at www.ICGtesting.com
Printed in the USA
BVIW12n0721230418
514158BV00001B/3